Kids Explore the Human Immune System

An Imaginative Guide to Human Biology for Kids — Made Fun and Easy

By: Z.B. Tucker

What is the human immune system?

The **immune system** is a multi-layer **defense system** that protects the human body from being attacked from microscopic invaders like bacteria, mold, viruses and fungi. These small invaders are always trying to attack and enter into your body each day. If they enter your body you can experience many different types of sickness and diseases.

What are the different protective layers of the human body?

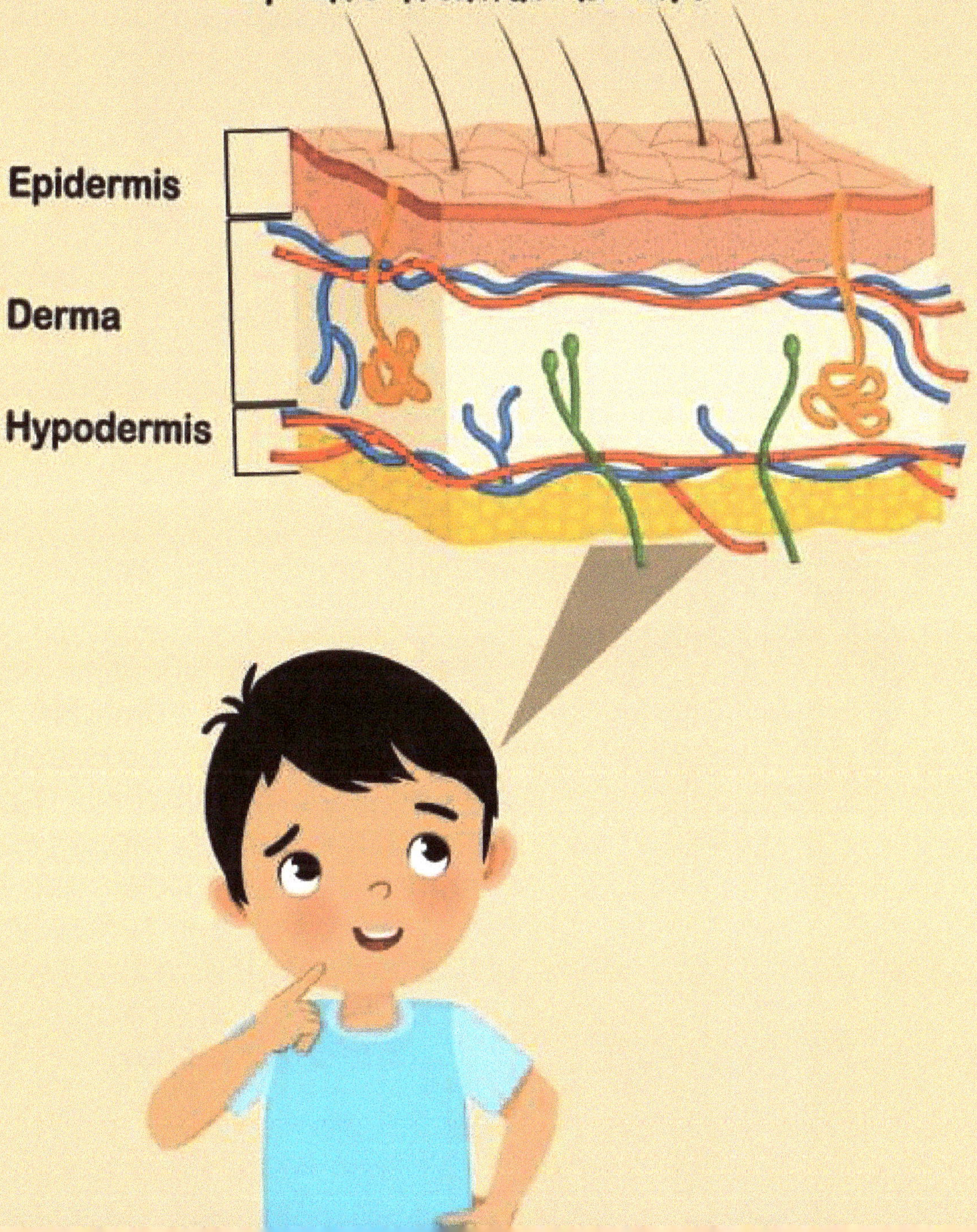

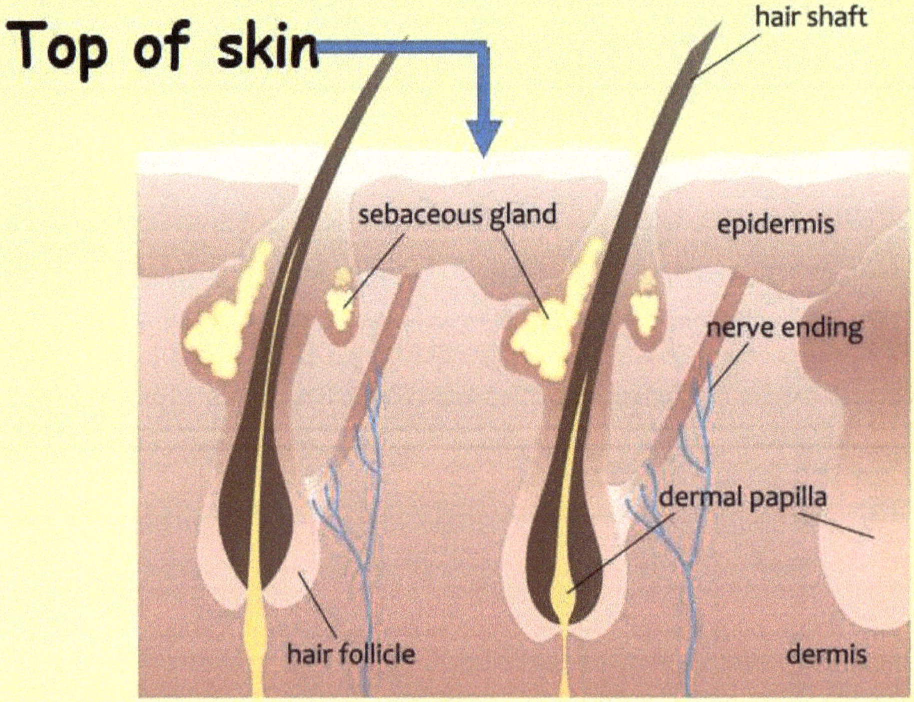

The first layer of the immune system begins with our **skin**. There is a thin outer protective layer on the skin called the **"Epi"** or **epithelial** layer. Although you cannot see this invisible coat, it is very important and serves as a **shield** to prevent anything from penetrating through the skin and entering the body.

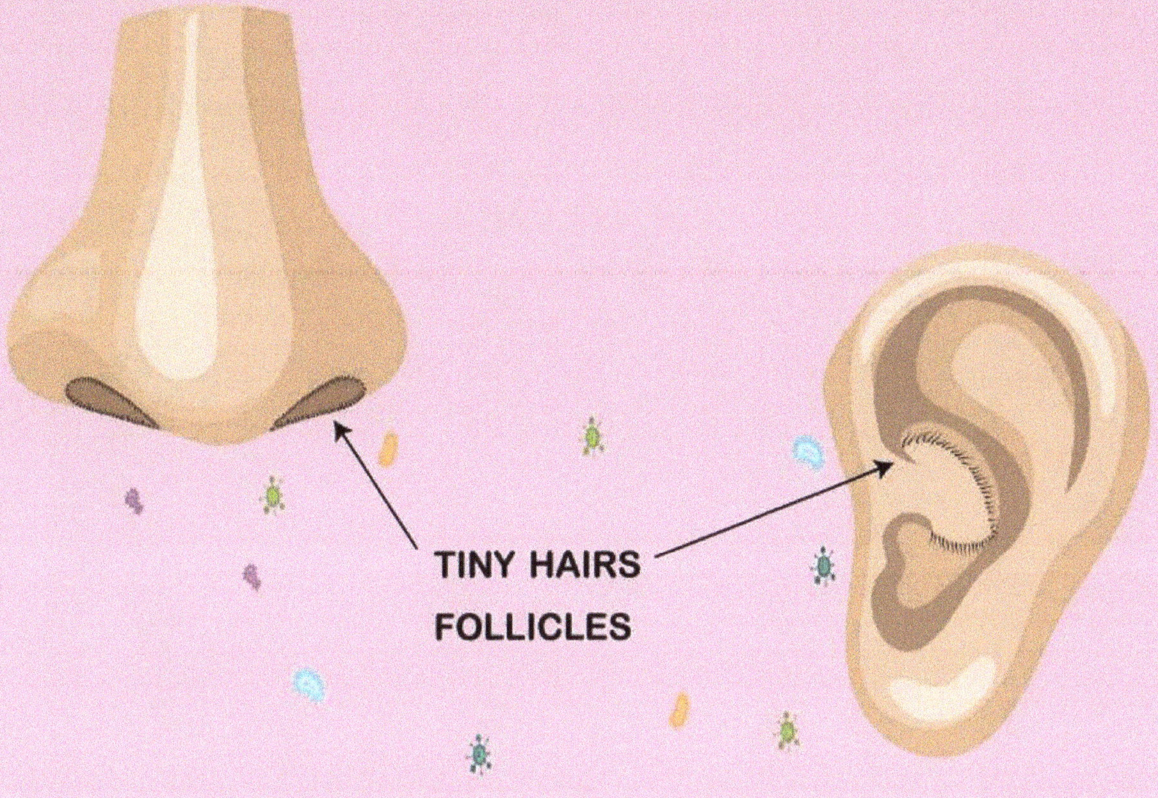

If the invaders can't enter through the skin, they may try to enter your body through your nose, mouth or ears. To prevent this, the nose and ears have tiny **hairs follicles** inside them to trap and catch anything from entering into the body through these openings.

SNEEZE **VOMITING** **COUGH**

If an invader enters the nose or mouth the body will often **sneeze**, **cough** or sometimes **vomit** to eject the intruders out. The strong rush of air will cause the harmful body to be kicked back outside.

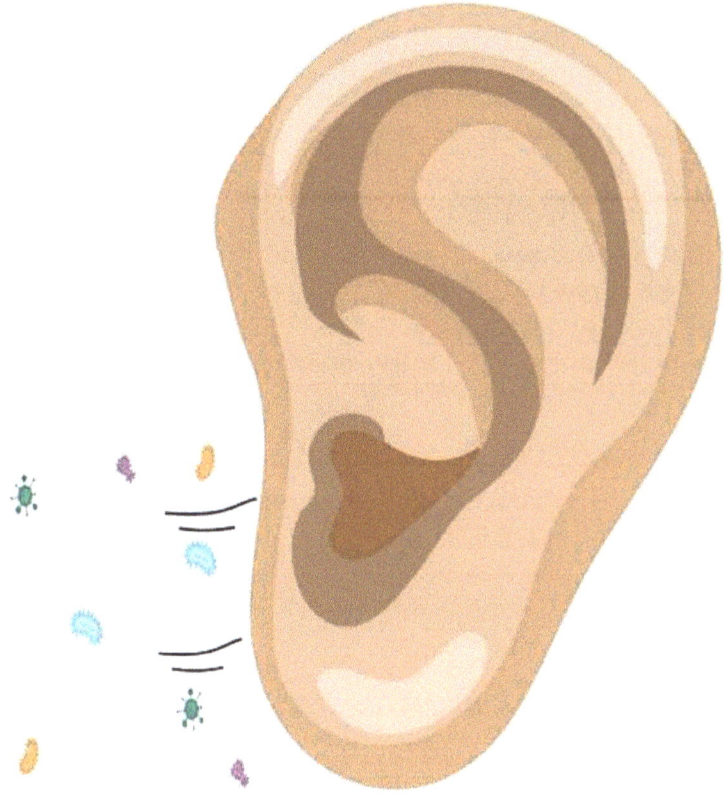

The ears also contain **wax**; an important sticky substance used to catch and trap any foreign invaders from going too far down inside them.

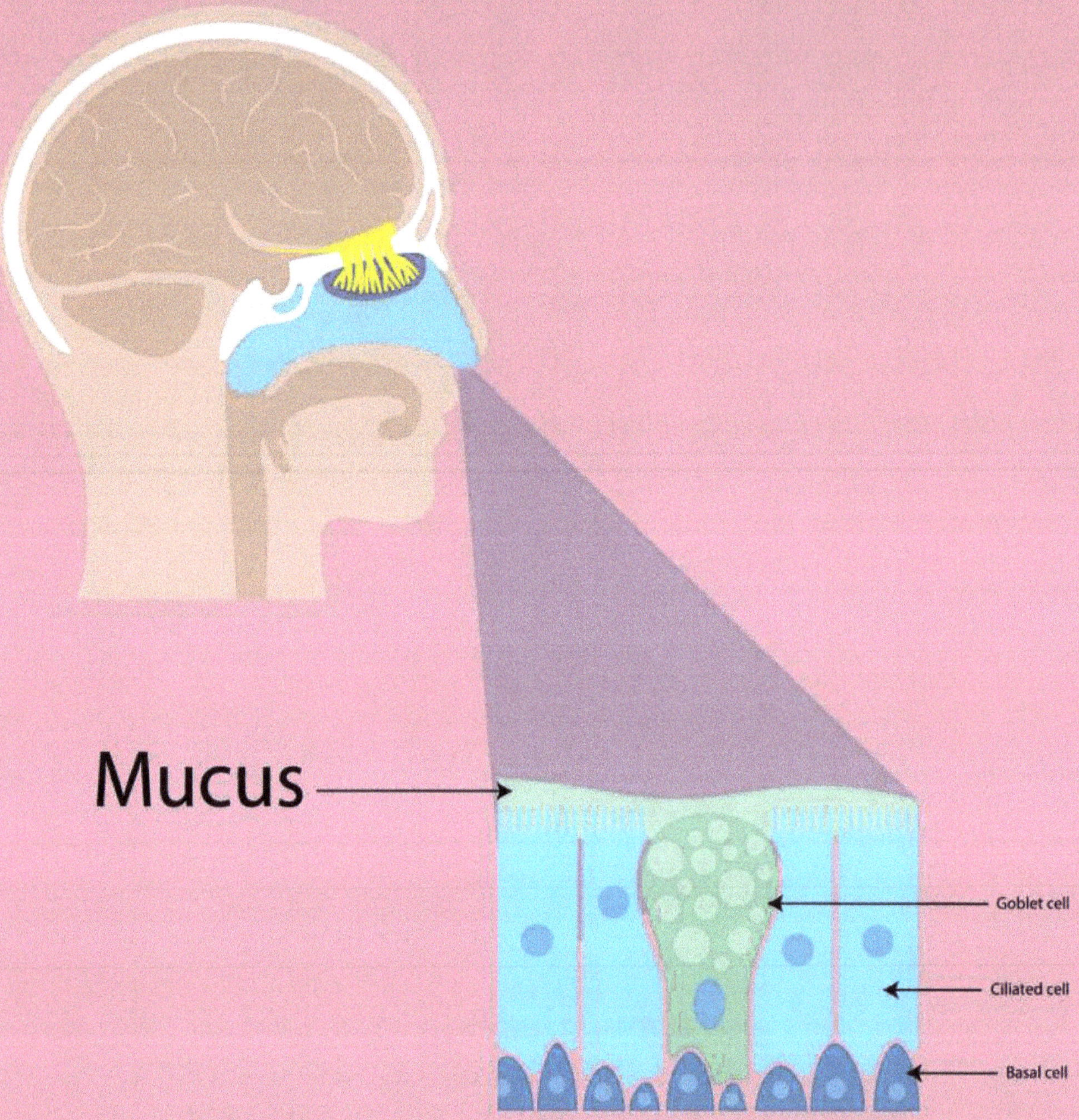

In addition to hair follicles, the throat and nose are also coated in a soft sticky substance known as **"Mucus"** or **mucosa** that can trap and catch any foreign invader from entering too far down into these openings.

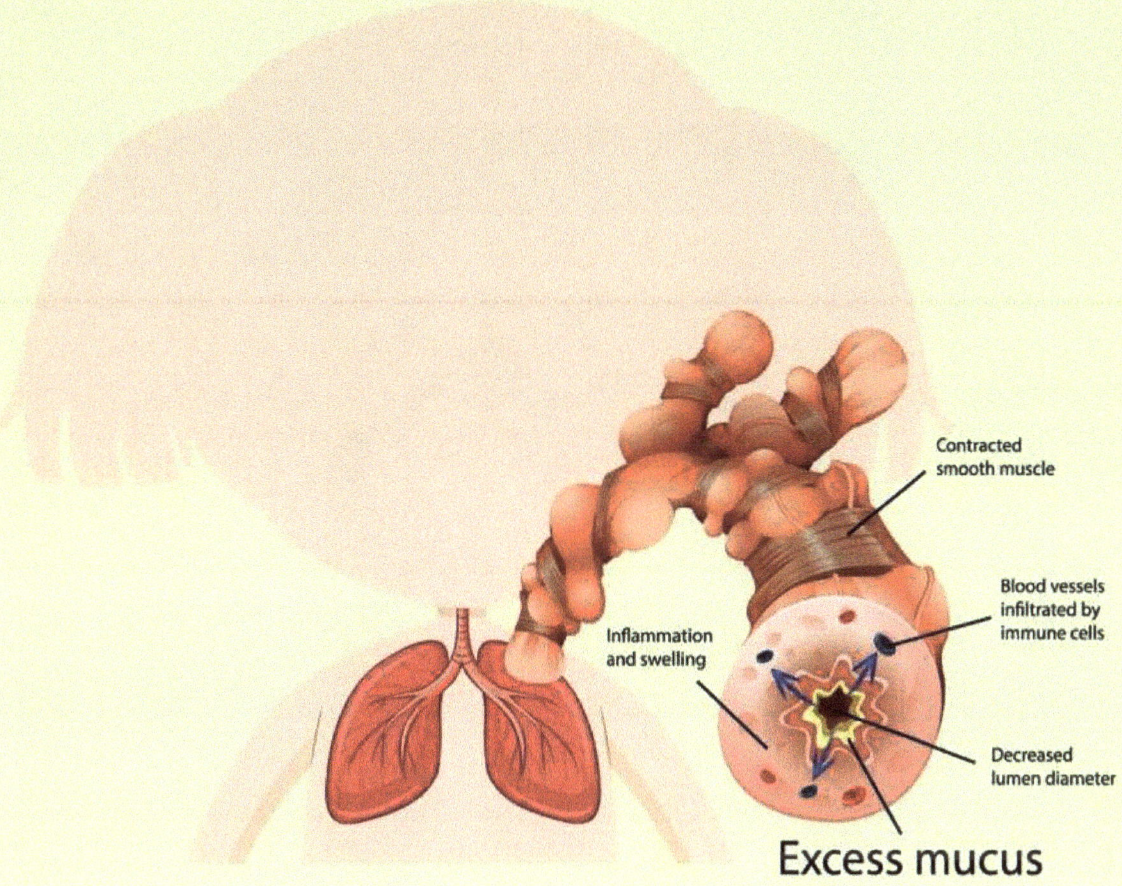

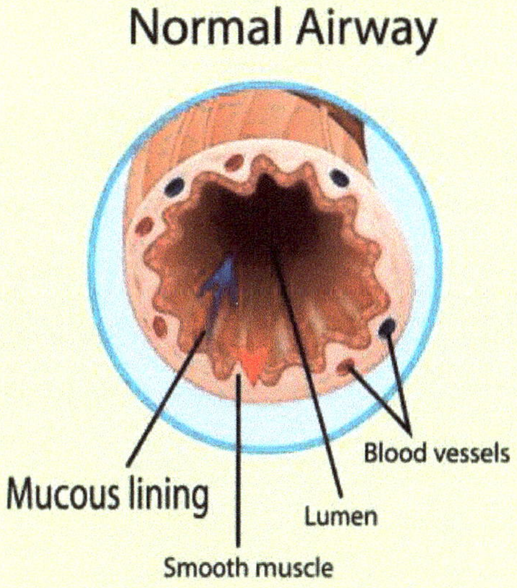

Mucosa is made inside the lungs to coat and protect the lungs against invaders. When you become sick the lungs produce a lot more mucus to ensure the lungs and throat are well protected.

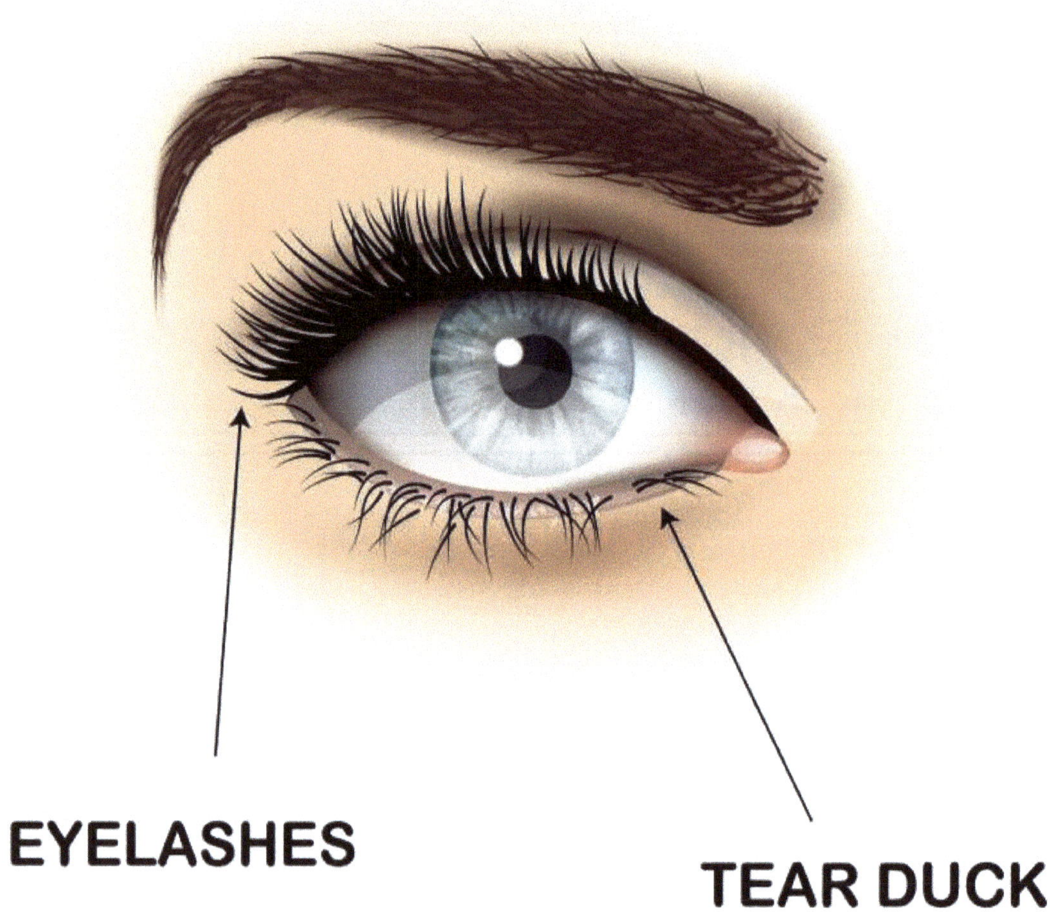

The eyes are protected by **eyelashes** and **eyelids** which help shield the eyes against foreign objects. The automatic **blinking** of the eyes is important for it reduces the eyes from being exposed for too long to the environment and keeps the eyes coated in **tears**.

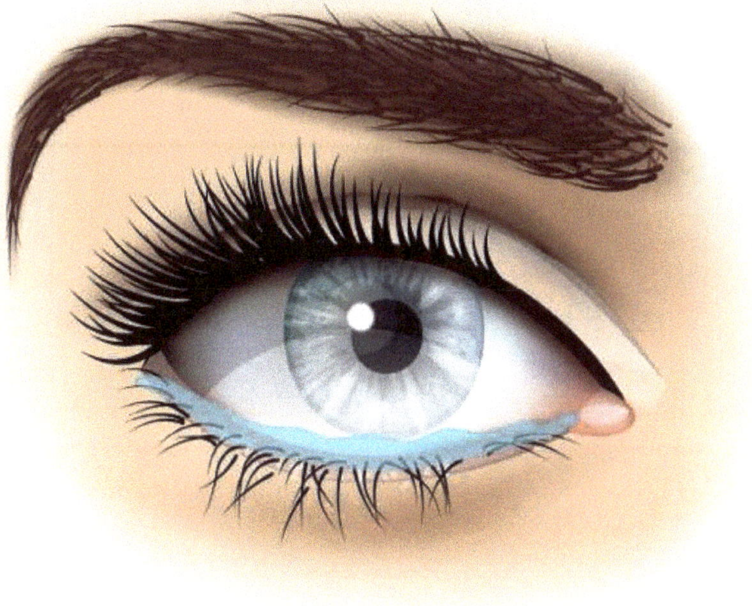

Tears are part of the human defense system. They consist of a salty substance that keeps the eyes moist and clean and can help sweep and drain away any foreign intruders that try to enter into the eye.

If something enters the eyes, or if the eye becomes too dry, it will create these **tears** on its own to flush out the invader. It may look like you're crying but it is only the eyes cleaning itself.

What happens if something gets pass all the outer protective defensive layers?

INFECTION

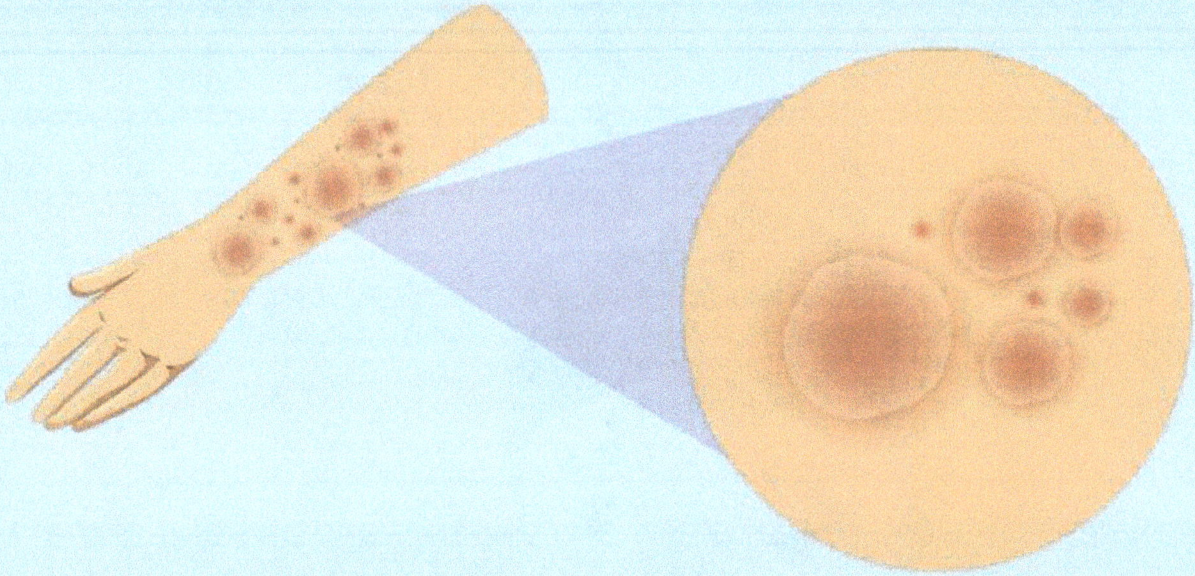

If a **foreign invader** is able to get pass the first protect layers of the body it will now go inside and will try to grow, multiply and spread throughout the interior of the body. The foreign body is often called an **infection** and will try to make your body into its new home.

FEVER

One of the first things the immune system will do is to increase your internal body temperature. Viruses and bacteria do not like hot environments and increasing the internal **temperature** of your body will help to stop and slow down their invasion. The temperature increase is often called a **fever** and can make you feel very hot and uncomfortable.

White blood cells

The second thing the immune system will do is send out **white blood cells** to fight the intruders and a **great war** will develop inside your body. These white bloods cells are made up of an army of "**Big Tanks**" called **Macrophages** that will eat many of the invaders trying to take over your body.

Your white blood cells eating diseases

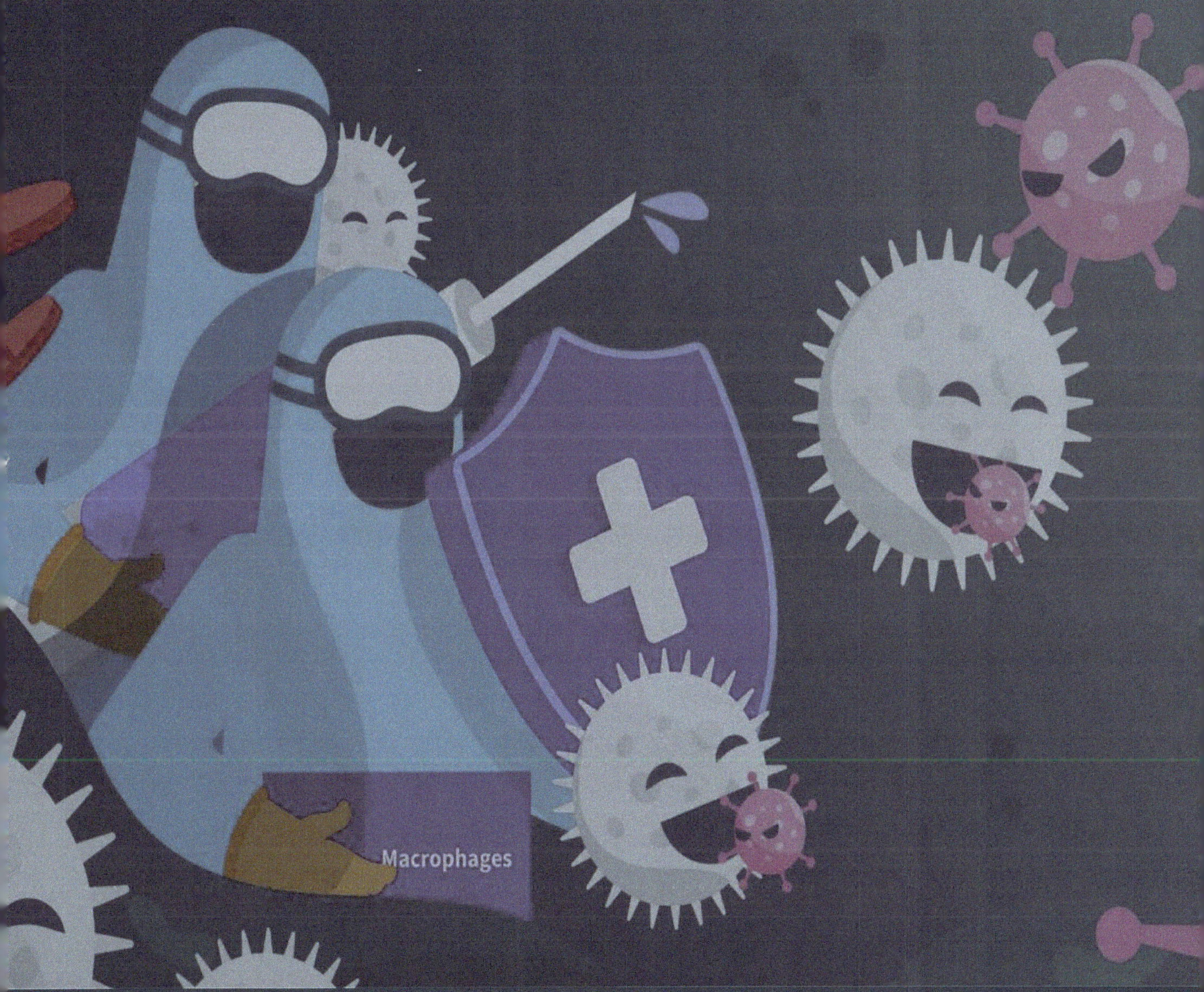

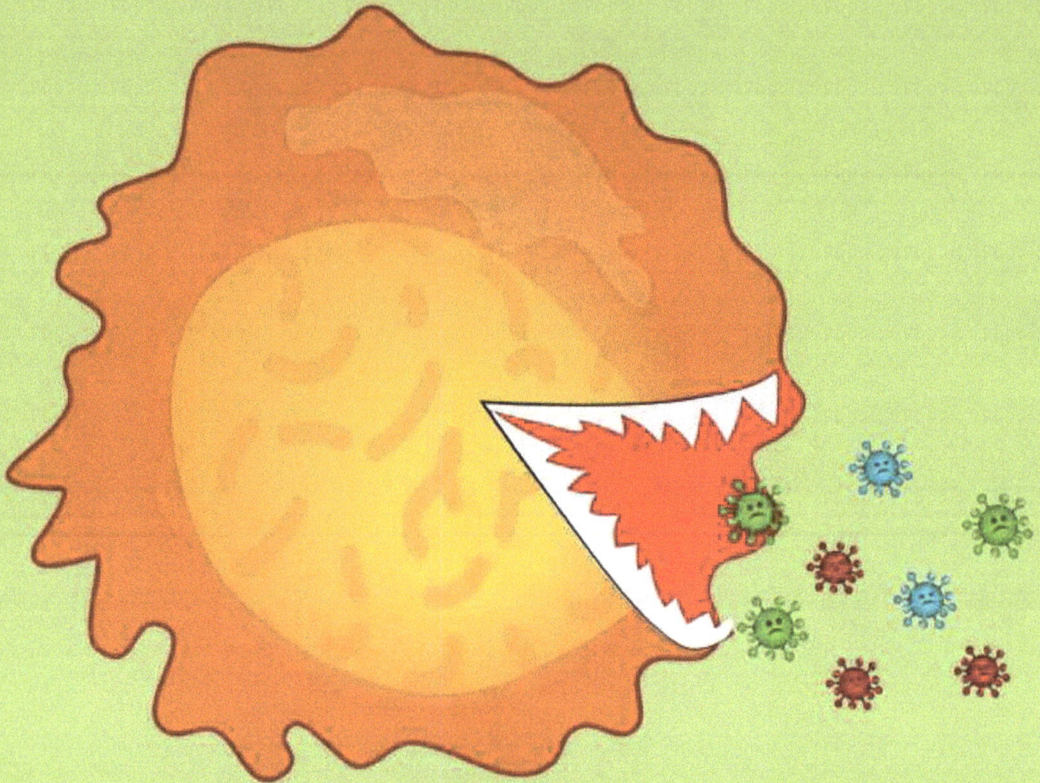

T cell

More white blood cells, called **T- Cells** will also arrive on the scene and help fight the intruders. **T-Cells** are like thousands of foot soldiers and their primary job is to help the **Macrophages** defeat the viruses and foreign invaders trying to start a new home in your body.

B cell

A third white blood cells called **B-cells** will soon arrive on the scene. Their job is not to fight but to investigate and analyze the invading foreign bodies. **B-cells** are looking for a permanent solution called **antibodies** to defeat the invaders once and for all.

B cell

T cell

White blood cells

These **antibodies** are like **super soldiers** and will be designed specifically to end the fight against that particular invader which is attacking you and will help to provide a long-term solution for defeating the intruders if they ever try to come back again.

B cell

T cell

White blood cells

These antibodies will provide your body with **immunity** against specific intruders.

ANTIBODIES

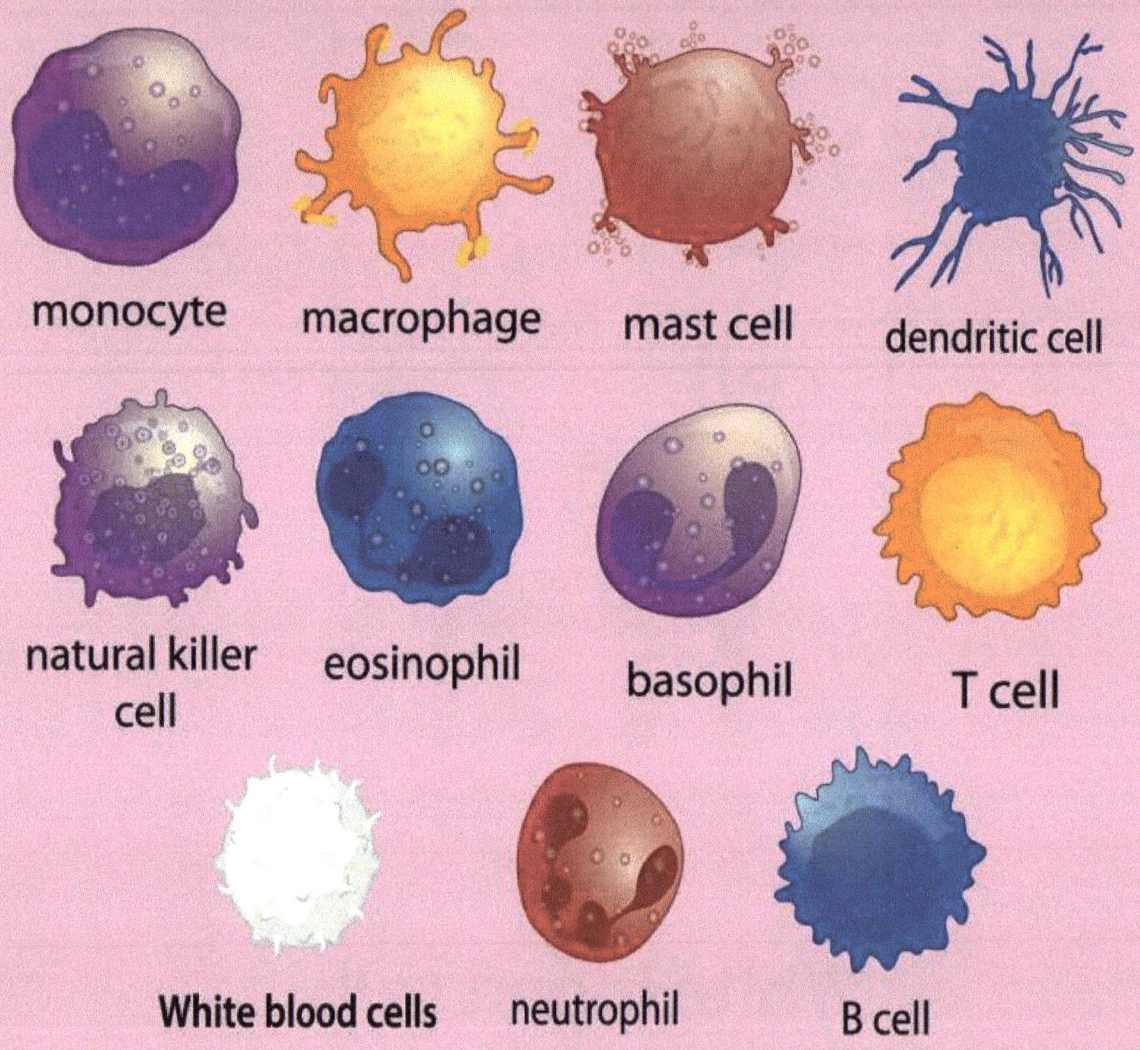

The immune system will keep all these different **antibodies** that it create stored inside you over your lifetime like a great library and they will be waiting if your body ever gets reinfected. The immune system will call on these super soldier antibodies to help protect you.

LEUKOCYTES

White blood cells called **Louie's** or **Leukocytes** are like your body's guardian angels. **Leukocytes** will continue watching and monitoring your entire body for any invaders through a system of roads and highways they travel through called the **lymphatic system**. Inside the **lymphatic system** these white blood cells act as **bodyguards**. They move about and will keep a constant eye out for any viruses and bacteria that may try to enter. **Leukocytes** are always watching.

IMMUNE SYSTEM

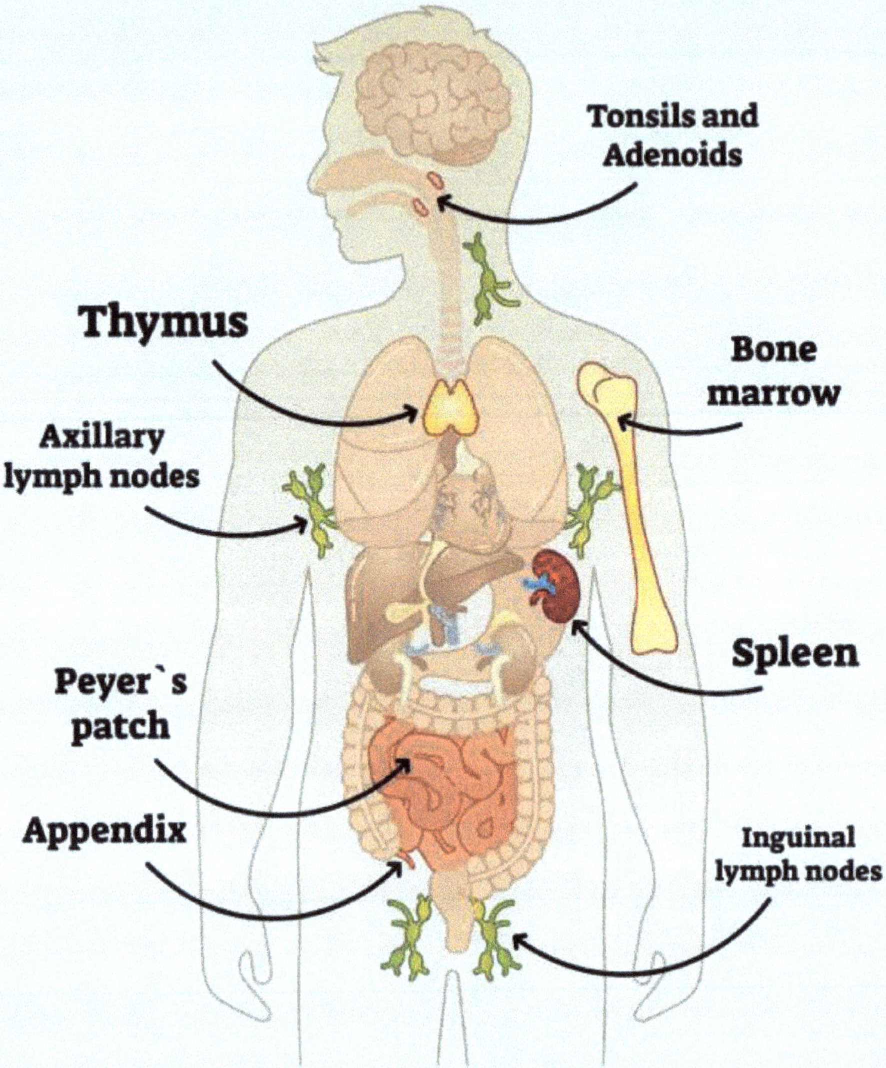

The Immune System is very complex and is designed with many things to protect you from getting sick. It is made up of a variety of defense systems both on the inside and outside of the body working day and night to help protect you. But sometimes things may go wrong with the immune system and it can stop protecting you. When this happens people often go to see an **immunologist**.

IMMUNOLOGIST

An **Immunologist** is a doctor who will diagnose and treat immune system problems like allergies or cancer. When your body's immune system does not function properly, it can lead to increase sickness and disease so an immunologist is an important doctor who can help solve these problems.Maybe you would like to become an immunologist one day to help people.

References

Bertok L, Chow D (2005). Bertok L, Chow D (eds.). Natural Immunity. ***5*** *(1st ed.). ISBN 978-0-44451-755-5.*

Janeway CA, Travers P, Walport M (2001). Immunobiology (5th ed.). Garland Science.

Sompayrac L (2019). *How the immune system works*. Hoboken, NJ: Wiley-Blackwell.

Author Bio

The author is an educator and a passionate lover of children. He received his Bachelor of Arts in Biology from Edgewood College, in Madison Wisconsin and a second B.A. in Healthcare Management from Southern Illinois University Carbondale where he also completed his M.A in Engineering. He has a wealth of valuable experience dealing with children of varying age ranges. The author does not only love children, but he understands that when they have the best form of education, it has a far-reaching effect on the wider society. As a result, he is on a mission to make sure that children learn as much as they can.

www.ingramcontent.com/pod-product-compliance
Lightning Source LLC
LaVergne TN
LVHW070900210926
848482LV00016B/980

* 9 7 9 8 9 9 7 3 6 6 1 3 1 *